はしがき

　この自動車ファイバースコープバックモニターは、従来のフェンダーミラー、ドアミラー、サイドミラーのように車体の横幅から外へはみ出さずに後方確認が容易にできるものであり、これにより悪天候による後方視界不良を回避し、また、接触の破損等を回避する為に構成されたものである。更に従来に於いては、左のドアミラー（サイドミラー）の後方を確認するには、顔を傾け、両目で見る必要があり、フェンダーミラーのように前方を見ながら確認できない点がマイナスであった。このマイナスを回避する為にファイバースコープからのバックモニターを運転席から前方の視界を見る範囲内のダッシュボードに設けることで快適な運転と安全を促進する。前記が初版のはしがきでありますが、バックモニターの本来の活用でもある駐車は、従来のフロント・リアカメラなどのモニターでは見られない独自の図形ライン基本パターンの表現で解説した。

Preface

This automobile fiberscope back monitor, It is constituted in order to be able to perform a back check easily, without overflowing the breadth of the body outside like the conventional fender mirror, a door mirror, and a rearview mirror, and for this to avoid the poor rearward visibility by bad weather and to avoid breakage of contact, etc.

Furthermore, in the former, in order to have checked the back of the left door mirror (rearview mirror), the point which cannot be checked while it is necessary to lean a face and to see by both eyes and the front is looked at like a fender mirror was minus.

In order to avoid this minus, comfortable operation and safety are promoted by preparing the back monitor from fiberscope in the dashboard within the limits which look at a front field of view from a driver's seat.

It was explained with an expression with the original figure line basic pattern which isn't judged from a monitor of a conventional front rear camera.

目 次

1、車庫入れ(イラスト解説)

(1) 図形ラインの幅寄せ、停止-- 9

(2) 停止後の操作---10

(3) 斜め前進停止後の操作

(4) バックの操作はラインに重ね合わせる---------------------------------11

(5) 左右同様

(6) 白線が消えている場合の操作--12

(7) 車庫入れ完了--13

2、前向き駐車(イラスト解説)

(1) ハンドル切り方の操作--14

(2) ラインに合わせてハンドルを切る--15

3、縦列駐車(イラスト解説)

(1) 幅寄せの白線に合わせる--16

(2) 角度に合わせてバック操作---17

2、英語解説

English description

1, Car warehousing (illustration explanation)

(1) ---19

Width draw and stop of a figure line

(2) ---20

Operation after a stop

(3) --20

Operation after a slanting advance stop

(4) --21

The back operation is piled up on a line.

(5)

Like left and right.

(6) --22

The operation when a white line becomes extinct

(7) --23

The end of car warehousing

2、The previous parking (illustration explanation)

(1) --24

Operation of the handle How to cut it

(2) --25

turn the steering wheel according to Rhine.

3、Column parking (illustration explanation)

(1) --26

It's added to a white line of pulling.

(2) --27

It's operated back according to the angle.

3、中国語解説

1、 有车库

　　⑴ --29

把图形线的宽度送到，停止

　　⑵ --30

停止後的操作

　　⑶

斜前進停止後的操作

　　⑷ --31

背部的操作為線疊合

　　⑸

左右同樣

　　⑹ --32

白線消失的情況的操作

　　⑺ --33

車庫盒完成

2、 面向前方停車

　　⑴ --34

方向盤終結方面的操作

　　⑵ --35

合起為線扭轉方向盤

3、 縦隊停車

(1) ---36

合起為盡量靠邊的白線

(2) ---37

合起為角度背部操作

4、ドイツ語解説

1, Das Parken Garage

(1) ---38

Halt der Zahl-Linie ziehend

(2) ---39

Operation nach dem Halt

(3)

Operation nach dem diagonalen Fortschritt-Halt

(4) ---40

Die Operation des Rückens erlegt es der Linie superauf

(5)

Das linke Recht ist dasselbe

(6) ---41

Operation, wenn eine weiße Linie verschwindet

(7) ---42

Das Parken Garage ist zu Ende

2、 Schicken Sie das Parken nach

   (1) --43

Die Operation des Steuerrades, wie man schneidet

   (2) --44

Ich drehe das Steuerrad in der Anpassung zur Linie

3、 Ich richte es vertikal und Park aus

   (1) --45

Ich vergleiche es mit einer weißen Linie des Ziehens

   (2) --46

Es wird der Rücken bedient, der zu einem Winkel fährt

1，
(1) 図形ラインの幅寄せ、停止

バックモニター車庫入れは、モニターを切り替えて、図形ライン、緑、青に幅寄せの白線に合わせて、運転操作を行えば、初心者でも上手に車庫入れができる。

バックの車庫入れについて、スペースの入り口の白線、中央付近に図形ラインの緑の位置を停止させる。

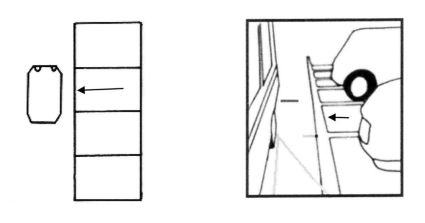

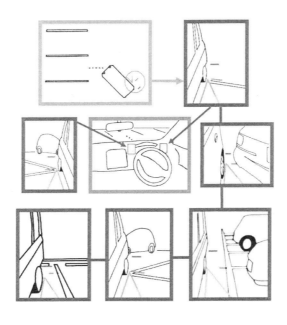

(2) 停止後の操作

　　停止後、ブレーキを踏みハンドルを左に目一杯切る。ブレーキペタルを解除し、約3メートル前進する。

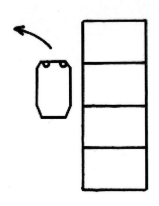

(3) 斜め前進停止後の操作

　　3メートル前進停止後、ハンドルを目一杯右に切り、ゆっくりバックする。

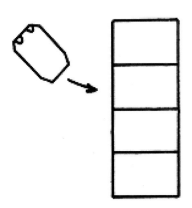

⑷ バックの操作はラインに重ね合わせる

バックモニターの図形ラインのV形の青、オレンジ又は黄色の角度に白線を重ね合わせながらバックする。

⑸ 左右同様

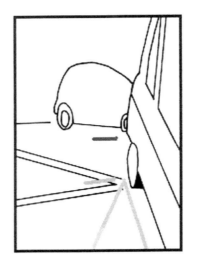

(6) 白線が消えている場合の操作

白線またはラインが消えて確認できない場合は、図形ラインの緑をバンパーに合わせて、V形ライン内にバンパーが入らないようにバックする。

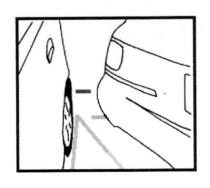

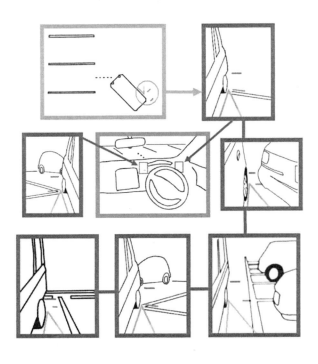

(7) 車庫入れ完了

　ハンドルを左回転させながら車が真っすぐになる時、図形ラインの赤、停止の位置を確認する。

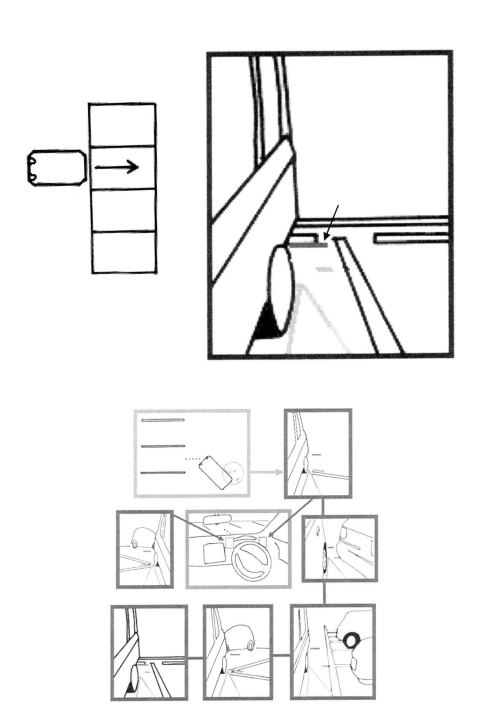

２、前向き駐車

(1) ハンドル切り方の操作

外側、内側の接触は、ハンドルの切り方が重要になっていますが、このバックモニターの図形ラインは、接触を避けるハンドルの切り方をガイドする。

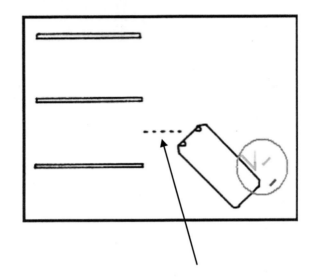

駐車スペースの中央に車を合わせる。

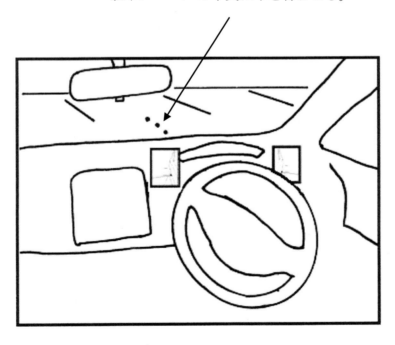

(2) ラインに合わせてハンドルを切る

　駐車スペースの入り口に平行する白線またはラインにバックモニターの図形ラインのＶ形の青、オレンジ又は黄色の角度に白線を重ね合わせる。
　ラインが無い場合は、イメージしたラインに合わせる。

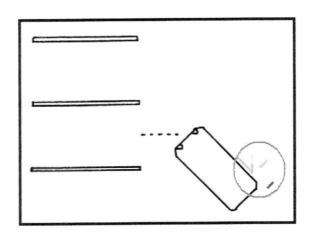

中央とＶ形が一致したらハンドルを切りながら駐車する。

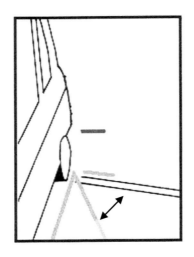

3、縦列駐車

(1) 幅寄せの白線に合わせる

バックモニター縦列駐車は、モニターの図形ライン、緑、青に幅寄せの白線に合わせて、運転操作を行えば、初心者でも上手に車庫入れができる。

バックの縦列駐車について、スペースの入り口の白線の図形ライン、緑の位置に車を停止させる。

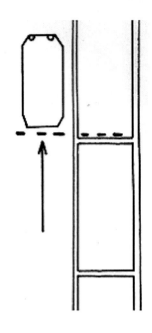

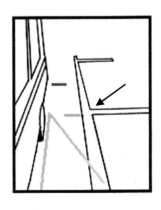

⑵ 角度に合わせてバック操作

左サイトのバックモニターの図形ラインのV形の青、オレンジ又は黄色の角度に白線を重ね合わせながらバックする。

図形ラインの赤、停止の位置を確認する。

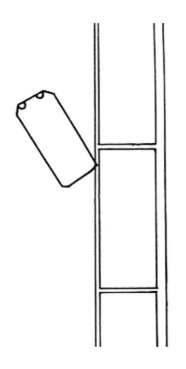
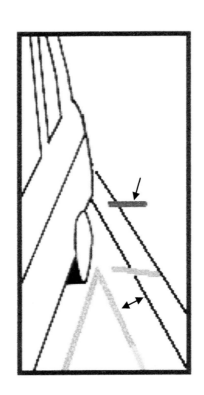

２，英語解説

When finding an unclear point by translation, please check it by a color illustration of the cover.

１，Car warehousing

⑴

When the back monitor depot insertion changes a monitor and does an operation according to the white line of pulling in a figure line, green and blue, even a beginner can do depot insertion well.

make them suspend the green location of the figure line around the white line and the center at the entrance of the space about the back depot insertion.

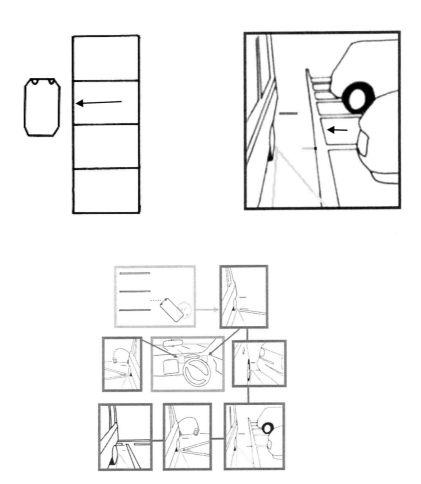

(2)

Operation after a stop

step on a brake after a stop and turn the steering wheel fully in the left. I release a brake petal and move ahead about 3 meters.

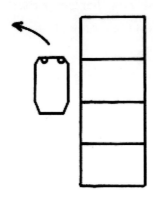

(3)

Operation after a slanting advance stop

turn 3 meters of steering wheel to the right fully after an advance stop, and it backs slowly.

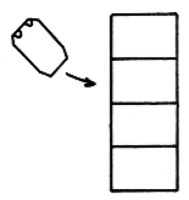

(4)
The back operation is piled up on a line.

It backs to blue of vee bottom, Orange or the yellow angle of a figure line of the back monitor while piling up a white line.

(5)
Like left and right.

(6)
The operation when a white line becomes extinct
When disappear, and the white line or the line can't be confirmed, it backs so that a bumper doesn't enter Midori of a figure line in the vee bottom line according to the bumper.

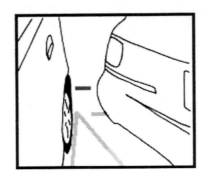

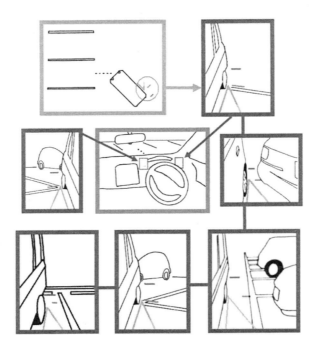

(7)
The end of car warehousing
When a car becomes straight though it makes them do the left spin of the handle, the red of a figure line and the location of the stop are confirmed.

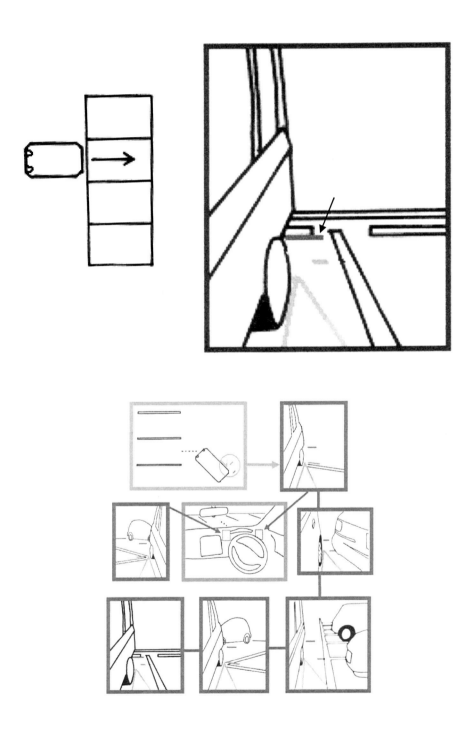

2、The previous parking (illustration explanation)
(1)
Operation of the handle How to cut it

How to cut it the handle becomes important to the inner contact outside, but a figure line of this back monitor guides how to cut it the handle from which contact is put aside.

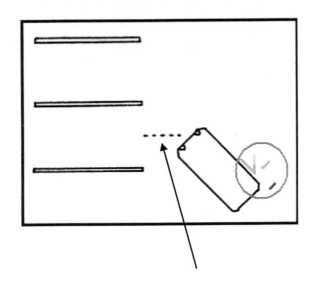

A car is added to the center of the parking space.

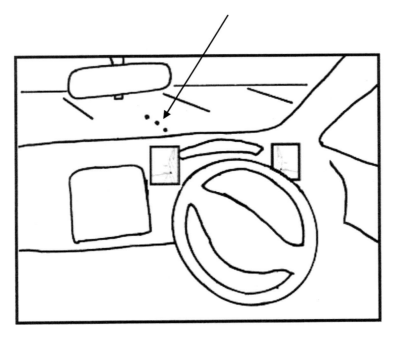

(2)

turn the steering wheel according to Rhine.

A white line is piled up on blue of vee bottom, orange or the yellow angle of a figure line of the back monitor in the white line or the line parallel makes an entrance of the parking space.

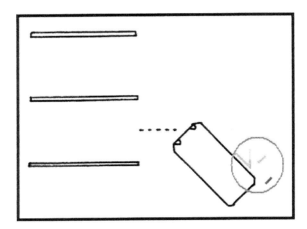

If vee bottom is parallel with the center, park while turning the steering wheel.

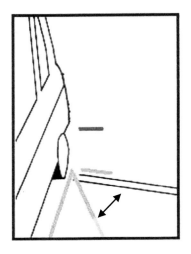

3、Column parking (illustration explanation)
   (1)
It's added to a white line of pulling.
When the back monitor column parking does an operation according to the white line of pulling in monitor's figure line, green and blue, even a beginner can do depot insertion well.
make a figure line of a white line and the green location at the entrance of the space stop a car about the back column parking.

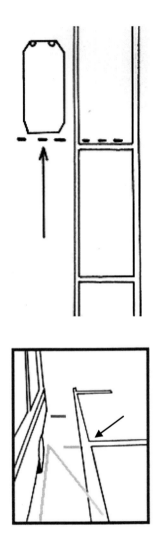

(2)
It's operated back according to the angle.
It backs to blue of vee bottom, Orange or the yellow angle of a figure line of the back monitor of the left site while piling up a white line.
The red of a figure line and the location of the stop are confirmed.

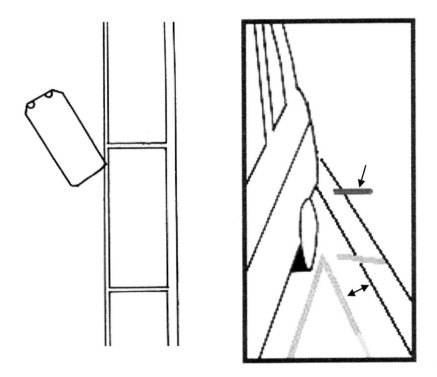

3、中国語解説

被发现时用翻译不明的点，用封面的颜色插图确认给。

1， 有车库

⑴

把图形线的宽度送到，停止

背部显示器车库盒，转换显示器，对图形线，绿，青合起为尽量靠边的白线，如果进行驾驶操作，连初学者也流利地车库盒作成。

关于背部的车库盒，到空间的入口的白线，中央附近使之停止图形线的绿的位置。

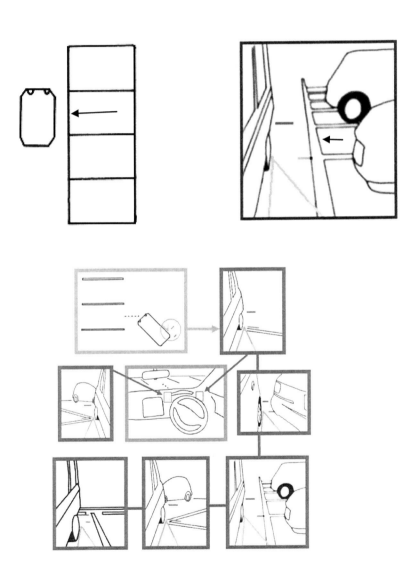

⑵ 停止後的操作

踏刹车，停止以后，向左最大限度地切方向盘。 解除刹车踏板，约 3 米前进。

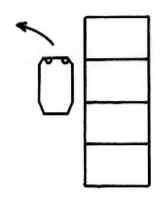

⑶ 斜前進停止後的操作

前進停止後 3 米，在眼一杯右切方向盤，慢慢地後退。

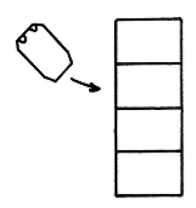

⑷

背部的操作為線疊合

背部顯示器的圖形線的 V 形的青,橘子又為黃色的角度一邊疊合白線一邊後退。

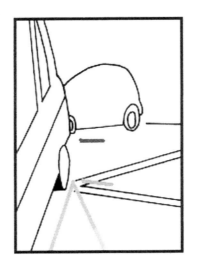

⑸

左右同樣

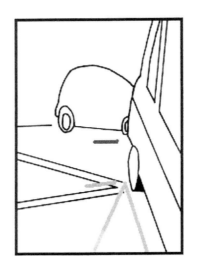

(6)
白線消失的情況的操作

白線或線消失如果確認，圖形線的綠合起保險杠，像 V 形線內中不引入保險杠一樣地後退。

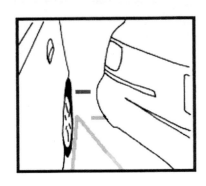

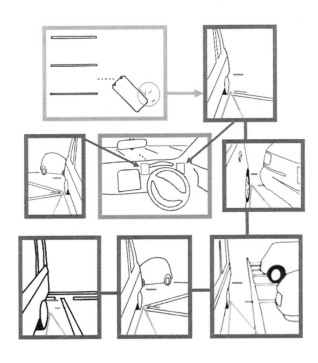

(7)
車庫盒完成

方向盤一邊使之左轉動一邊確認車變得直時，圖形線的紅，停止的位置。

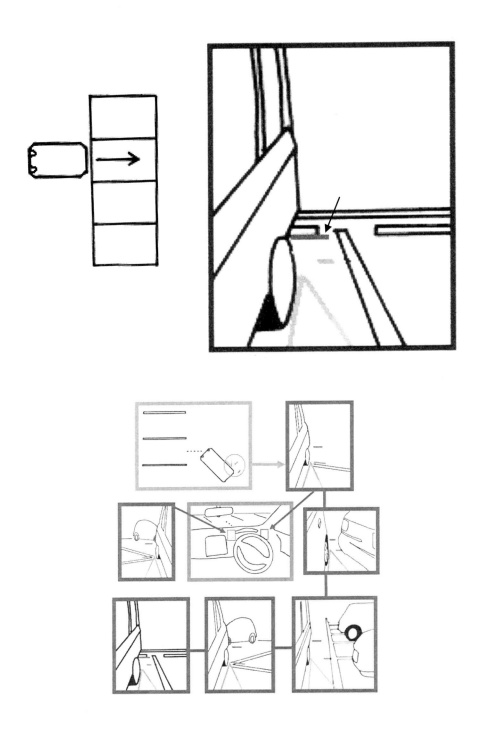

2、 面向前方停車

(1)

方向盤終結方面的操作

外側，內側的接觸，方向盤切方法變得重要，不過，這個 bakkumoni 的圖形線，做嚮導避開接觸的方向盤切方法。

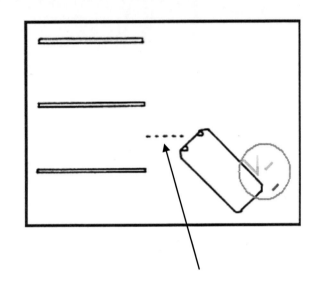

向車位中央合起車。

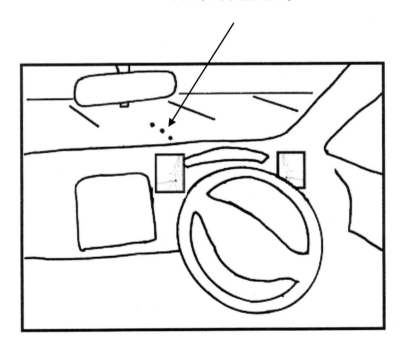

⑵

合起為線扭轉方向盤

為車位的入口並行的白線或為線背部顯示器的圖形線的 V 形的青，橘子又為黃色的角度疊合白線。

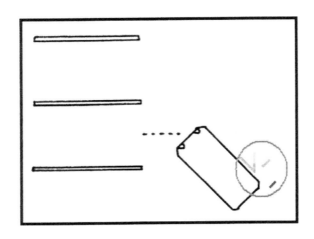

如果中央和 V 形相符一邊扭轉方向盤一邊停車。

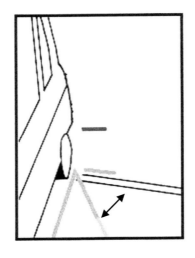

3、 縱隊停車

⑴
合起為盡量靠邊的白線

背部顯示器縱隊停車，對顯示器的圖形線，綠，青合起為盡量靠邊的白線，如果進行駕駛操作，連初學者也流利地車庫盒作成。

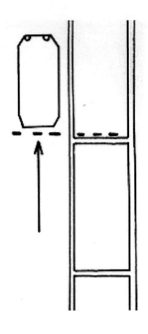

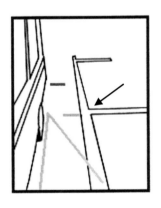

⑵

合起為角度背部操作

左網站的背部顯示器的圖形線的 V 形的青，橘子又為黃色的角度一邊疊合白線一邊後退，圖形線的紅，確認停止的位置。

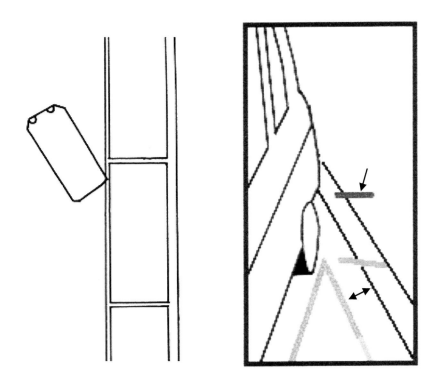

4、ドイツ語解説

Als irgendwelche Fragen in der Übersetzung entdeckt wurden, bestätigen Sie es bitte mit dem colorplate des Deckels.

1, Das Parken Garage

(1)

Halt der Zahl-Linie ziehend

Wenn das Parken einen Schirm ändert und Kontrolle in der Anpassung zu einer Linie, dem Grün der Zahl, einer weißen Linie des Blaus durchführt, kann sogar ein Anfänger es gut abstellen.

Dort stellt ab und hört die grüne Position der Zahl-Linie in der Nähe von einer weißen Linie, dem Zentrum des Eingangs des Raums auf.

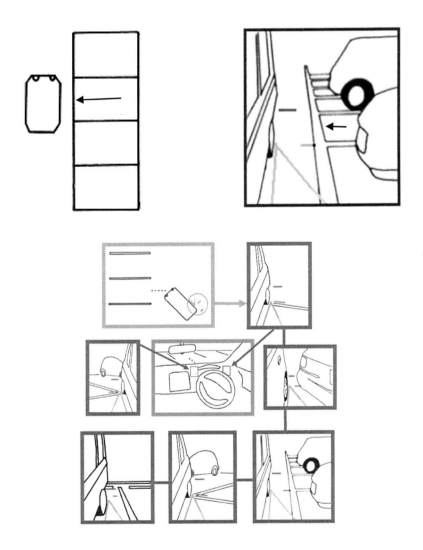

(2)
Operation nach dem Halt
Ich gehe auf Bremsen und nach einem Halt, drehe das Steuerrad bei der vollen Druckwelle nach links. Ich entferne Bremse-Pedal und etwa 3 Meter Fortschritt.

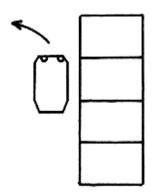

(3)
Operation nach dem diagonalen Fortschritt-Halt
Nach 3 Meter des Fortschritt-Halts drehe ich das Steuerrad zum vollen Recht und gehe langsam zurück.

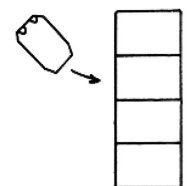

(4)

Die Operation des Rückens erlegt es der Linie superauf

Ich gehe zurück, während ich eine weiße Linie dem Blau, einem Orange der V-Gestalt der Zahl-Linie des Zurückmonitors oder eines gelben Winkels superauferlege.

(5)

Das linke Recht ist dasselbe

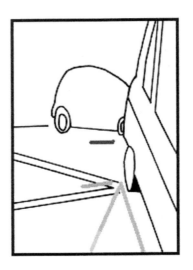

(6)
Operation, wenn eine weiße Linie verschwindet

Wenn eine weiße Linie oder eine Linie verschwinden und es nicht bestätigen können, vergleiche ich das Grün der Zahl-Linie mit einer Stoßstange und gehe zurück, so dass eine Stoßstange in die V-Gestalt-Linie nicht eingeht.

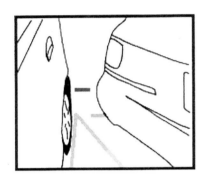

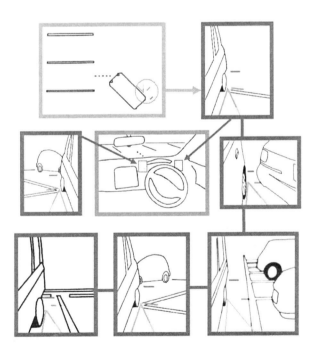

(7)
Das Parken Garage ist zu Ende
Ich bestätige das Rot der Zahl-Linie, die Position des Halts, während ich Sie ein Steuerrad gegen den Uhrzeigersinn rotieren lassen lasse, wenn ein Auto gerade wird.

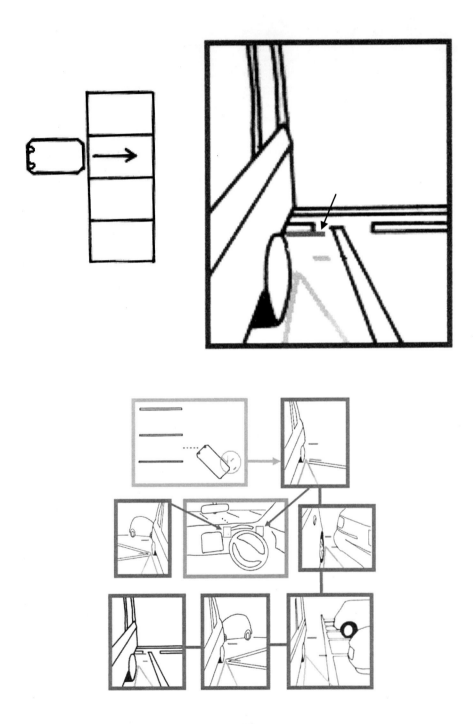

2、 Schicken Sie das Parken nach
 (1)
Die Operation des Steuerrades, wie man schneidet

Wie zur Kürzung des Steuerrades wichtig wird, aber, bezüglich der Außenseite, des Innenkontakts, führt die Zahl-Linie dieses Zurückmonitors wie zur Kürzung des Steuerrad-Vermeiden-Kontakts.

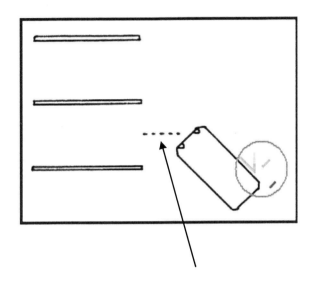

Ich vergleiche ein Auto mit dem Zentrum des Parkplatzes.

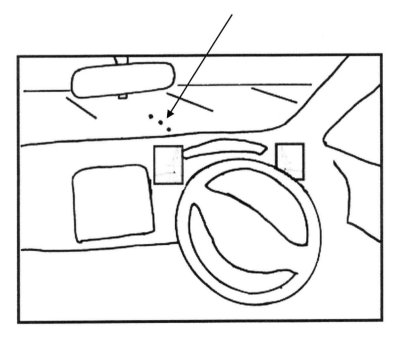

(2)

Ich drehe das Steuerrad in der Anpassung zur Linie

Ich habe eine weiße Linie oben auf einander zum Blau, einem Orange der V-Gestalt der Zahl-Linie des Zurückmonitors oder eines gelben Winkels auf einer weißen Linie in der Parallele mit dem Eingang des Parkplatzes oder der Linie gestellt.

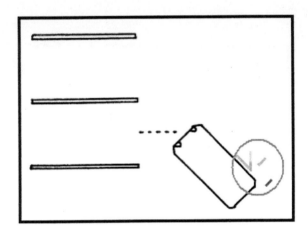

Ich parke, während ich das Steuerrad drehe, wenn V-Gestalt mit dem Zentrum harmoniert.

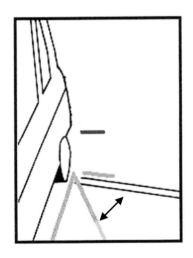

3、 Ich richte es vertikal und Park aus
 (1)
Ich vergleiche es mit einer weißen Linie des Ziehens
Wenn die Zurückmonitor-Säule parkend es zu einer weißen Linie des Ziehens in einer Zahl-Linie, grün, das Blau des Monitors verschmilzt und Kontrolle durchführt, gibt es einen Werkstatt-Fall sogar dem Anfänger gut.
Über das Säulenparken des Rückens höre ich ein Auto an der Zahl-Linie der weißen Linie des Eingangs des Raums, der grünen Position auf.

(2)

Es wird der Rücken bedient, der zu einem Winkel fährt

Ich gehe zurück, während ich eine weiße Linie oben auf einander zum Blau, einem Orange der V-Gestalt der Zahl-Linie des Zurückmonitors der linken Seite oder eines gelben Winkels stelle.

Ich bestätige das Rot der Zahl-Linie, die Position des Halts.

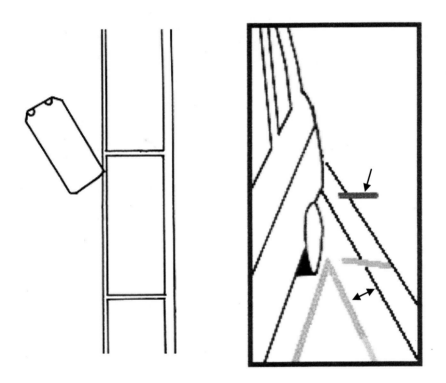

あとがき

　本書のバックモニターの活用範囲は、初版でも解説の如く、著者からの資料は出願当時の平成23年頃からであり、自動車、オートバイ、電動車椅子などの研究著作資料で主に安全性の研究であった。この資料の過程で総合的に著作権企画を行っていたものであり、今後、当初の著作資料に基づいて、追加出版を行う。

　また、2005年に創作された統制、ＡＰＩＣによる社会問題、人間並のペット保険登録、ＩＤカード（保険証・診察券）を発行によるシステムの流れの企画書・マニュアルなどの著作権・不正競争防止法による出版を発行予定。

　その他、著者は1992年以前から車庫付き中高層建物の研究もなされ、観覧車・エレベータ式などでない昇降方法で、安全・安価にできる構成であり、その構成内容を独創的表現で解説、その解説文献を明細書に引用された場合は、著作権侵害に抵触する可能性がある。これらの技術の解説表現による出版を今後、予定している。

　更に医療部門では、アミノ酸などを含んだ癌の治療薬や難病の研究をしており、著者は薬学の専門家でもあることから製薬会社から注目されている特効薬の文献の著作権出版も今後、予定している。

　本書のバックモニターに表示される図形ラインは基本パターンであり、パターンには数種類あり、これを文章の著作権で表現すれば「駐車を行う場合、モニターを駐車用の表示に切り替える。Ａパターンの図形ラインは極端に狭い駐車場、Ｂパターンの図形ラインは大型車などである。」これらの構成をイラスト、図形ライン、使用方法を著作権、表現で解説したものであり、更にこの企画の内容は不正競争防止法に鑑みたものである。

自動車ファイバースコープバックモニター　図形ラインで車庫入れ　前向き駐車　縦列駐車簡単

定価（本体1,500円＋税）

───

２０１５年（平成２７年）５月１５日発行

No.

発行所　IDF（INVENTION DEVLOPMENT FEDERATION）
　　　　発明開発連合会®
メール　03-3498@idf-0751.com　www.idf-0751.com
電話　03-3498-0751㈹
150-8691　渋谷郵便局私書箱第２５８号
発行人　ましば寿一
著作権企画　IDF発明開発(連)
Printed in Japan
著者　牧野　真一 ©
　　　（まきのしんいち）

初版、２０１４年（平成２６年）３月５日発行に記載できなかった原稿の追加発行

───

本書の一部または全部を無断で複写、複製、転載、データーファイル化することを禁じています。

It forbids a copy, a duplicate, reproduction, and forming a data file for some or all of this book without notice.